DATE DUE

AP 26 '97			
JY 1 2 '97			
OC 6 '97			
OC 07 '99			
JA 31 '01			
10/19/02			

DEMCO 38-297

SCIENCE TRICKS

Laurence B. White, Jr.

drawings by

Marc Tolon Brown

Addison-Wesley

Books by Laurence B. White, Jr.

Science Games
Science Puzzles
Science Toys
Science Tricks
Investigating Science with Coins
Investigating Science with Nails
Investigating Science with Paper
Investigating Science with Rubber Bands
So You Want to be a Magician?
The Great Mysto . . . That's You

An Addisonian Press Book

To Dave . . .
who guarantees every experiment
because they worked for him

Addison-Wesley Publishing Company, Inc.
Reading, Massachusetts 01867
Printed in the United States of America
First Printing

WZ/WZ 08611 12/75

Library of Congress Cataloging in Publication Data

White, Laurence B
Science tricks.

SUMMARY: Simple science tricks and experiments for the very young.
"An Addisonian Press book."
1. Science—Experiments—Juvenile literature.
2. Scientific recreations—Juvenile literature.
[1. Science—Experiments. 2. Scientific recreations]
I. Title.
Q163.W49 500 74-854
ISBN 0-201-08611-5

Tricks are to Try

- Make the end of your finger vanish!
- Bounce a real egg on the table!
- Pour water in a pail four feet away!

Are these magic tricks?

Yes? No! They are all science tricks for you.

Science tricks sometimes look like magic . . .

. . . and they are . . .

. . . until you understand how and why they work!

How can you understand them?

By doing them!

. . . . Have fun!

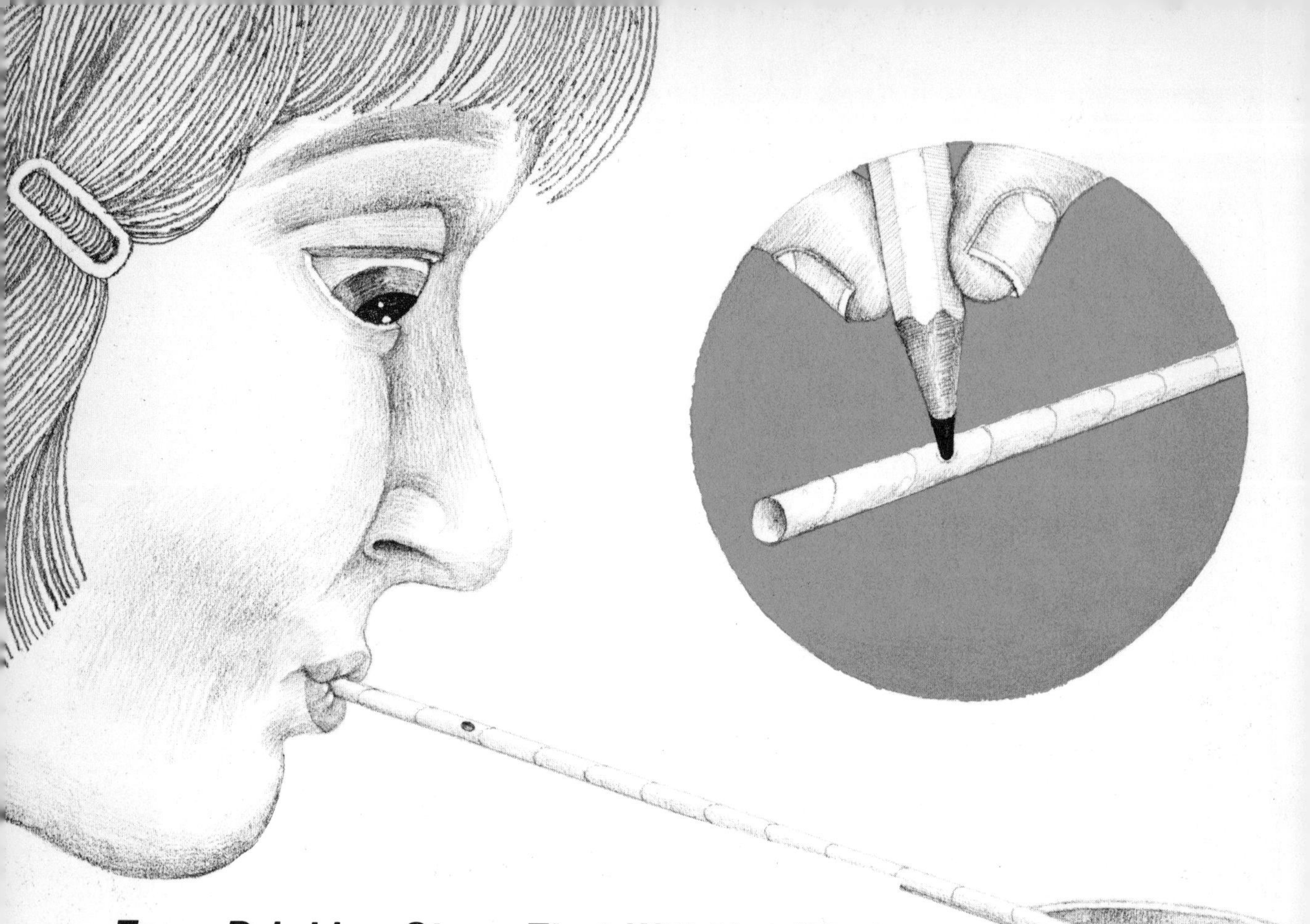

Try a Drinking Straw That Will Not Work

Fill a paper cup with water.
Sip some up with a drinking straw.
Now make a little hole in the straw.
Now try to sip water with it.
You will not be able to.
Air will go in the hole.
You will just get a drink of air.

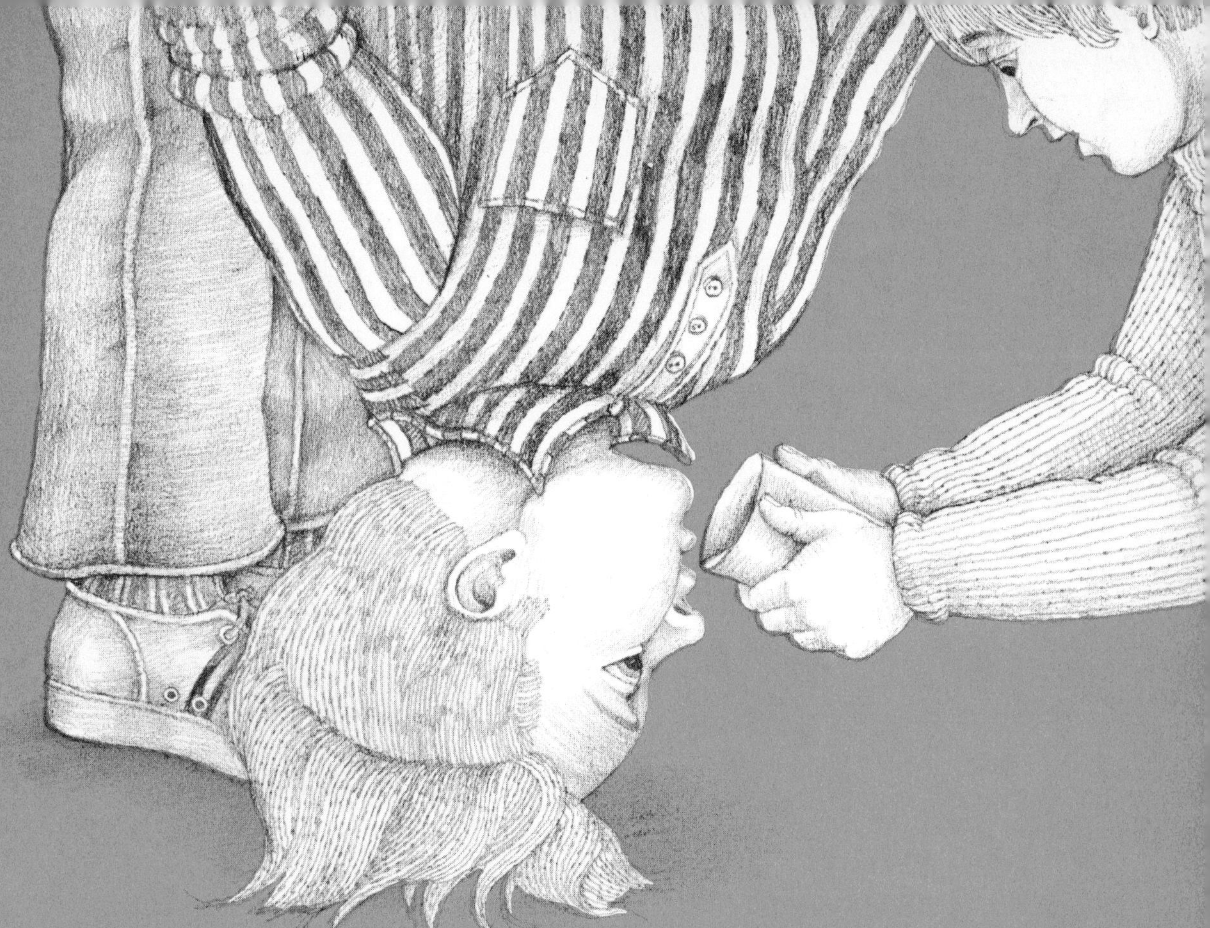

Have a Drink Upside Down

Have a friend hold you upside down.
Try drinking water from a paper cup.
You have lots of muscles.
Muscles make your body move.
They even help you eat and drink.
Muscles in your throat move water down . . . or up.

Have You Ever Heard a Fly Walking?

Catch a fly.
Put it in a paper cup.
Cover the cup with paper.
Hold the paper on with a rubber band.
Go where it is very quiet.
Hold the cup against your ear.
The cup is like a drum.
It makes sounds louder.
Stomp, stomp, stomp, stomp.
You will hear the fly walking.

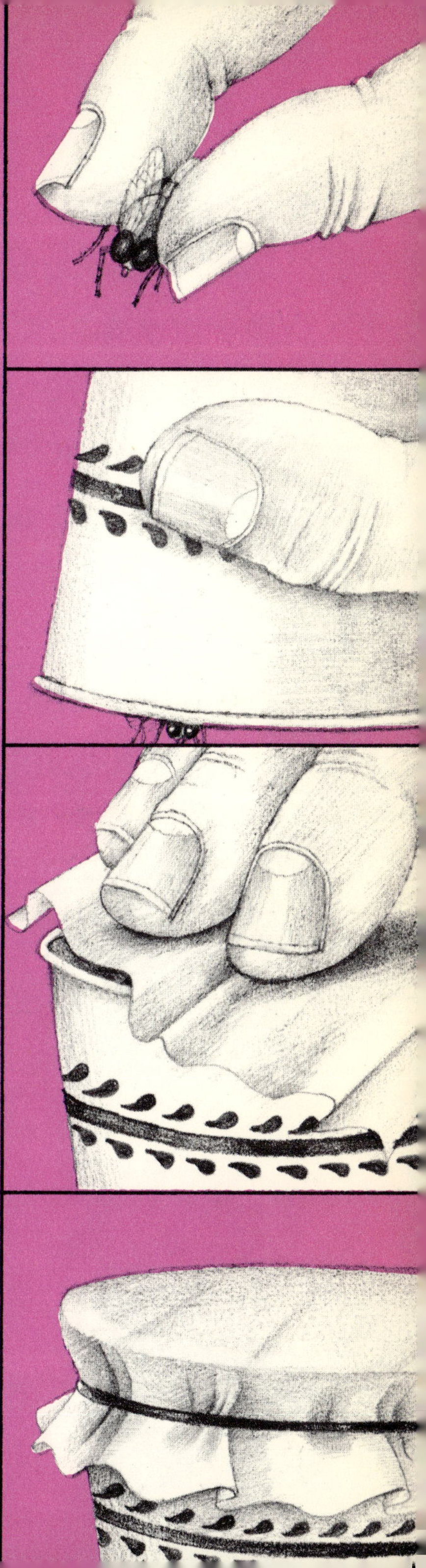

Hear the Sounds in a Paper Cup

There are sounds in a paper cup.
Very tiny sounds.
Hold a cup very close to your ear.
Put your finger in your other ear.
Ask everyone around you to be quiet.
Now listen!
You will hear sounds like wind or ocean waves.
Hold up glasses, pitchers and bottles.
Again in a quiet room . . .
Listen!
They all have sounds in them.
Sounds are everywhere.

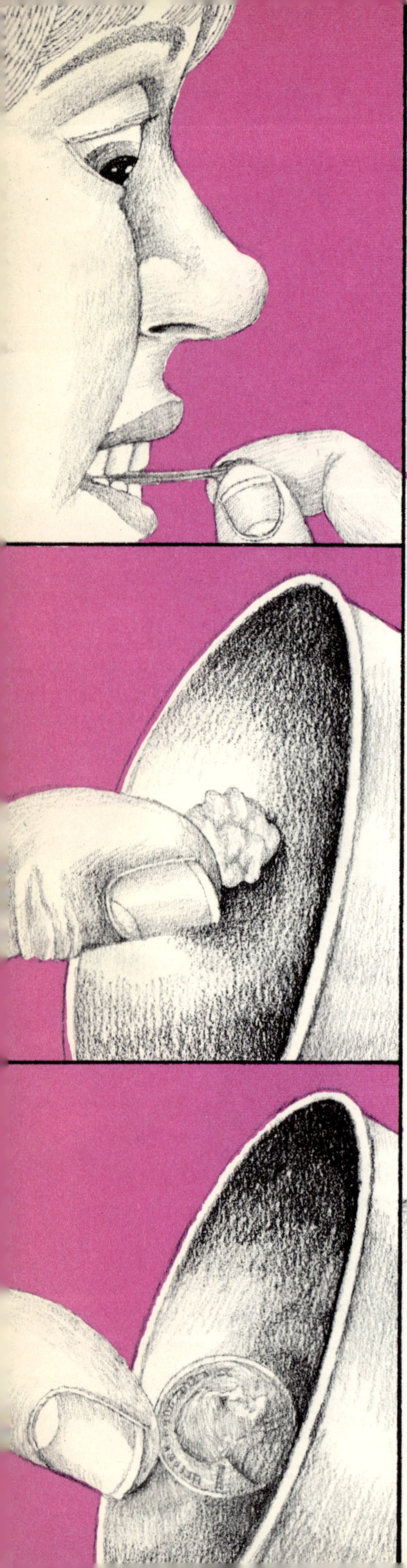

Make a Cup Move by Itself

Stick some chewing gum in a paper cup.
And, stick a coin on the gum.
Put the cup on the table with the coin up.
Let go.
The cup zooms back and forth.
The coin inside makes the cup move.

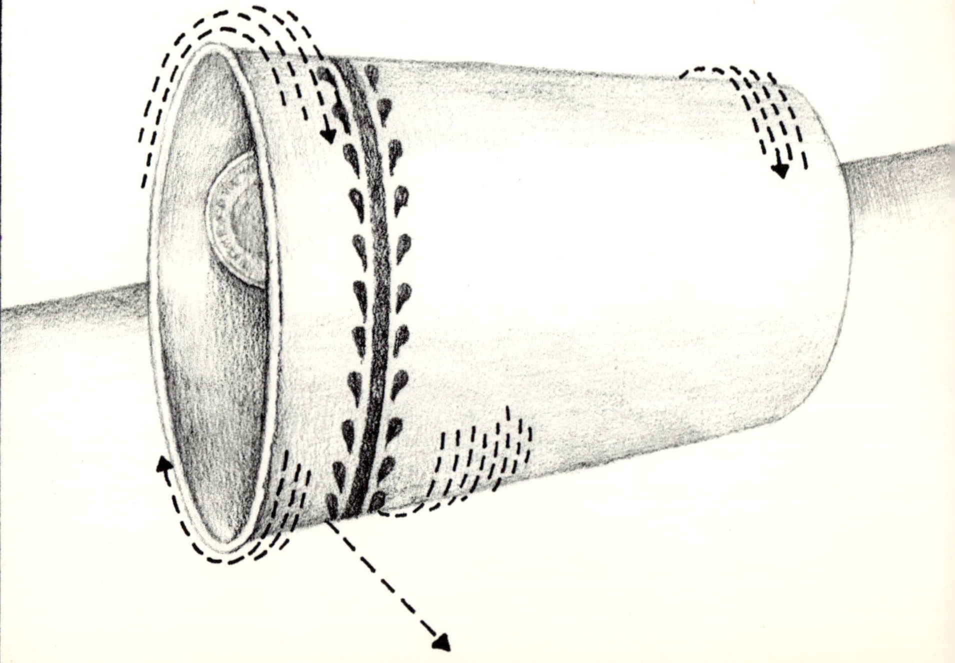

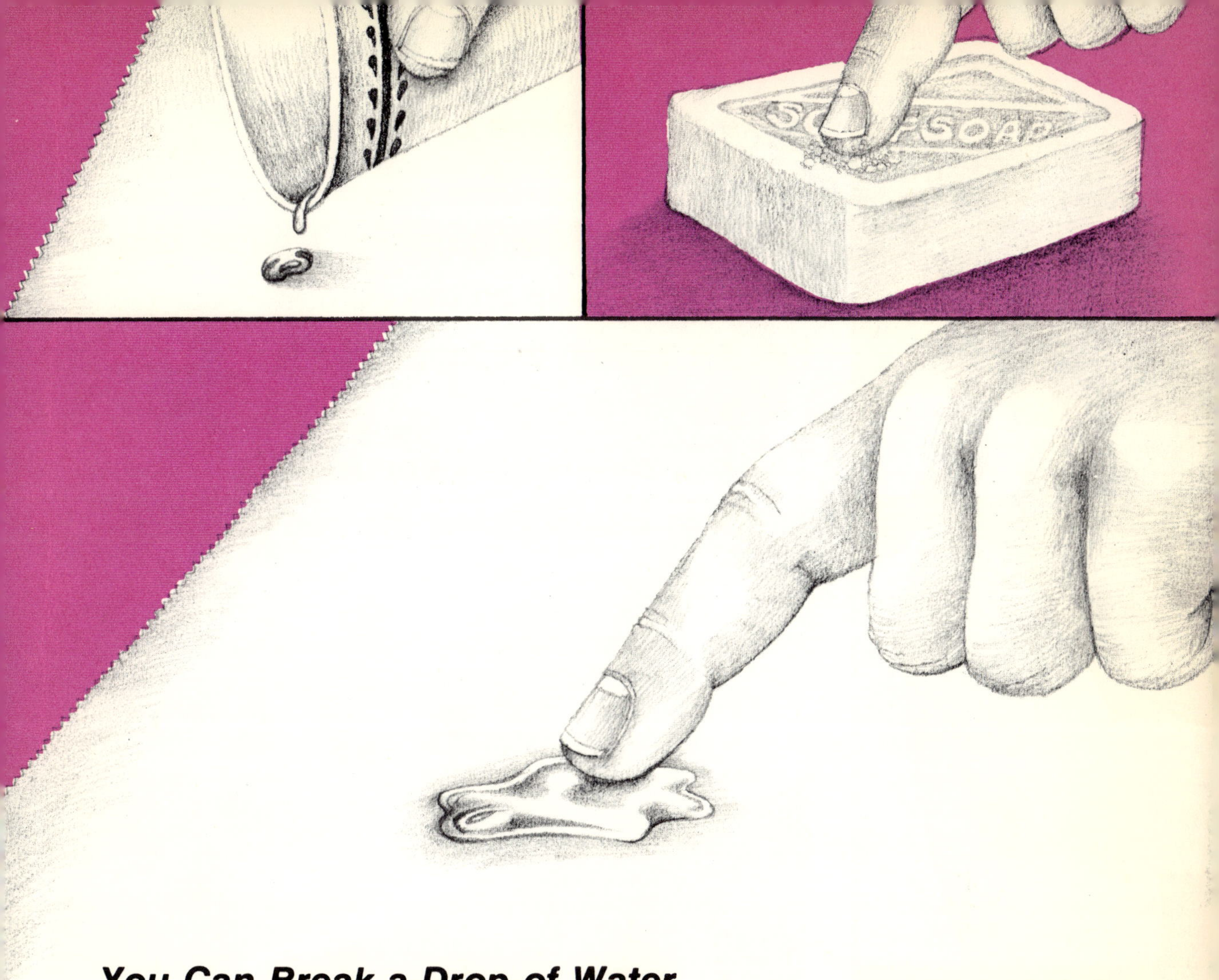

You Can Break a Drop of Water

Pour a big drop of water on a piece of wax paper.
Pour carefully, all in one place.
Rub your finger over a bar of soap.
Touch your soapy finger to the drop of water.
Like magic, soap makes the water drop break.

You stand here . . . .

Pour Water with a Broom

. . . and pour water in a pail over here.
How? With a broom.
Hold it by the brush end.
Tip it down a little.
Pour water on the handle.
Some water falls off.
Some sticks to the broom and runs down to the pail.
Why? Because water is sticky.
Some water sticks to your hands when you wash.
Some water sticks to your clothes when it rains.

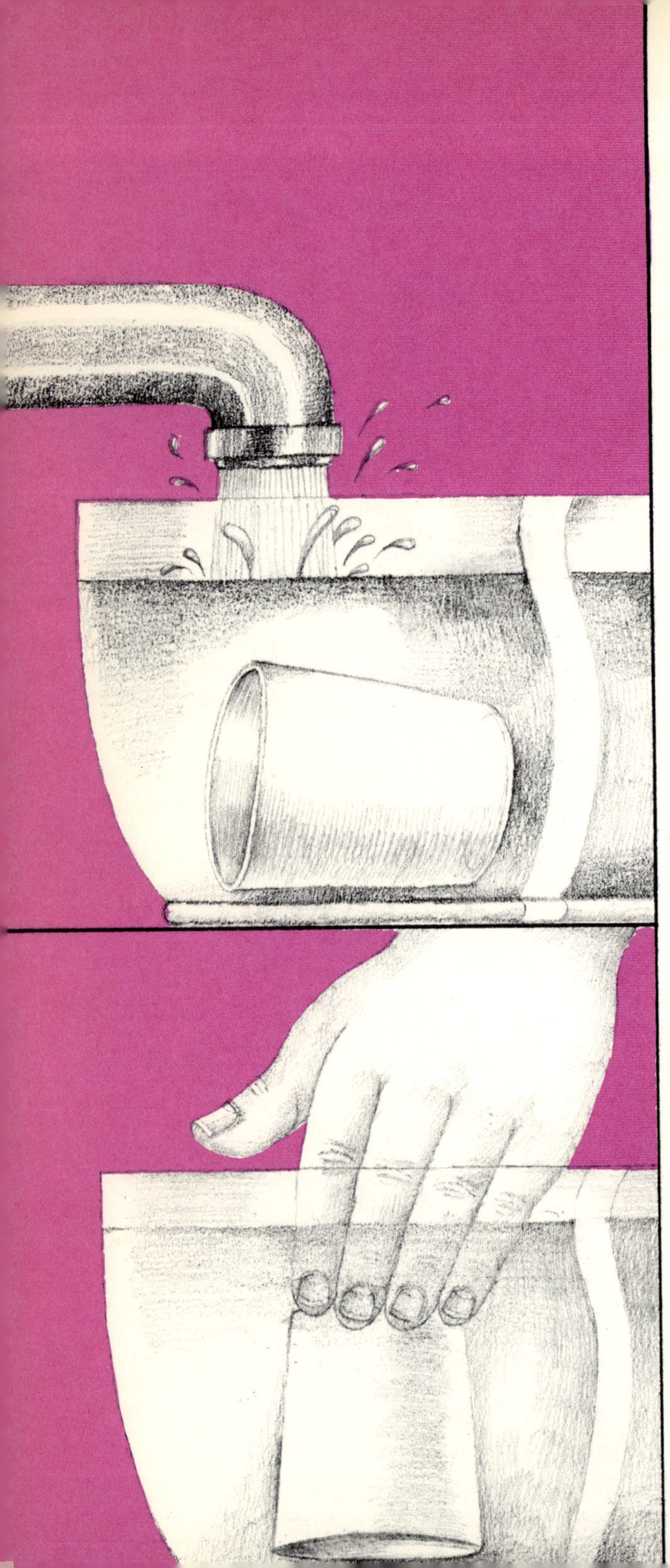

Lift an Upside Down Glass of Water

Fill a big bowl with water.
Sink a glass in the water.
Reach in and lift the glass.
Lift it upside down.
Keep the open end
under water.
The water will not run out.
Air cannot go into the glass.
Air must go in for the water
to run out.
Lift the glass higher.
Let some air in.
Splash!

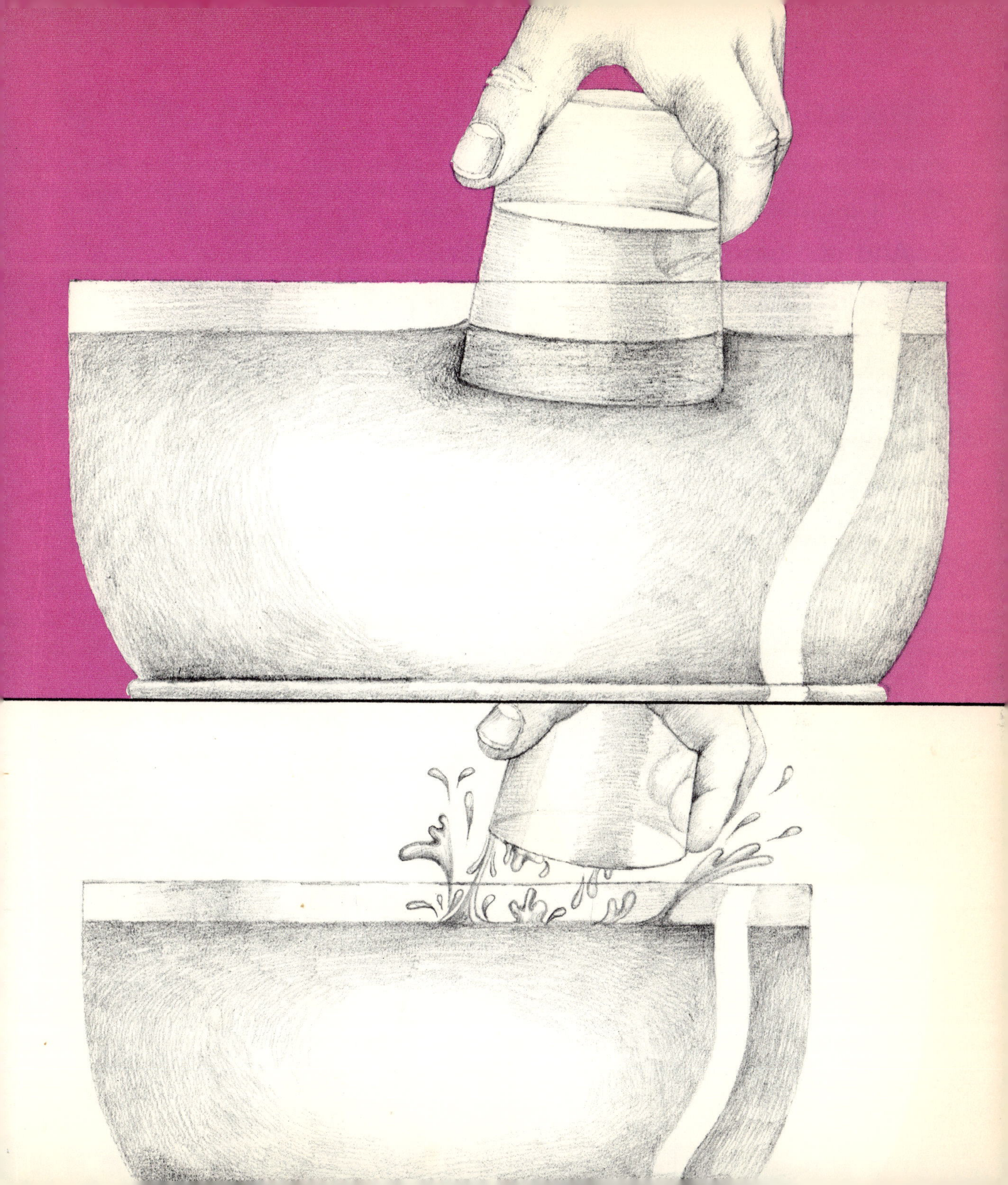

Mix a Glass of Bubbles

Fill a glass half full of water.
Add two big spoonfuls of vinegar.
Stir in a spoonful of dishwashing liquid.
Mix in a big spoonful of baking soda.
Woosh!
The glass is full of bubbles.
Vinegar and baking soda make a gas.
The gas blows tiny bubbles of soap.

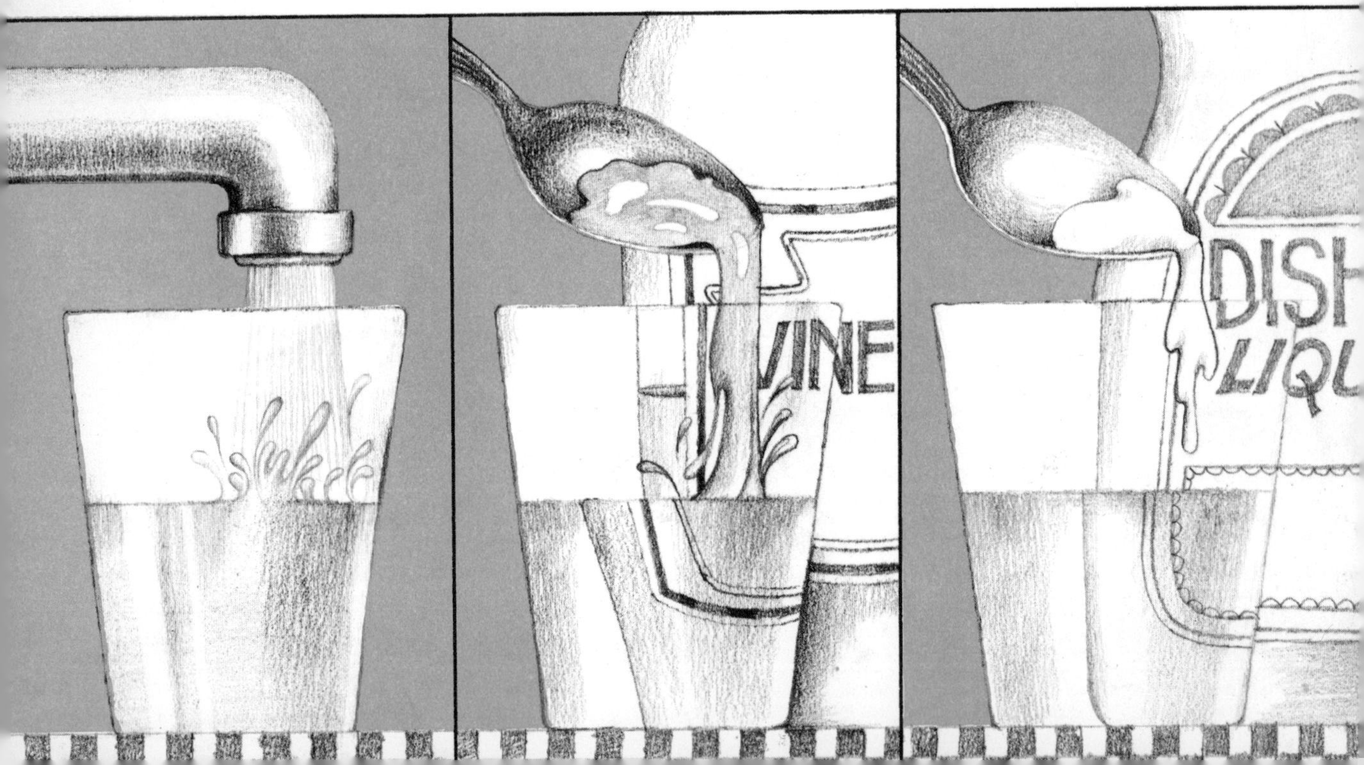

BAKING
POWDER

Make an Egg that Bounces

Ask Mom to hard boil an egg for you.

Put it in a drinking glass half full of water.

Then fill the glass with vinegar.

Leave it until tomorrow.

It will look the same.

Feel it.

Bounce it! (Do it gently.)

Vinegar makes the hard shell soft.

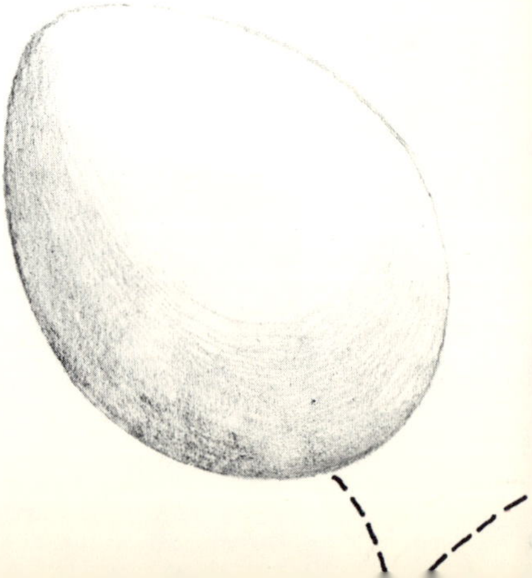

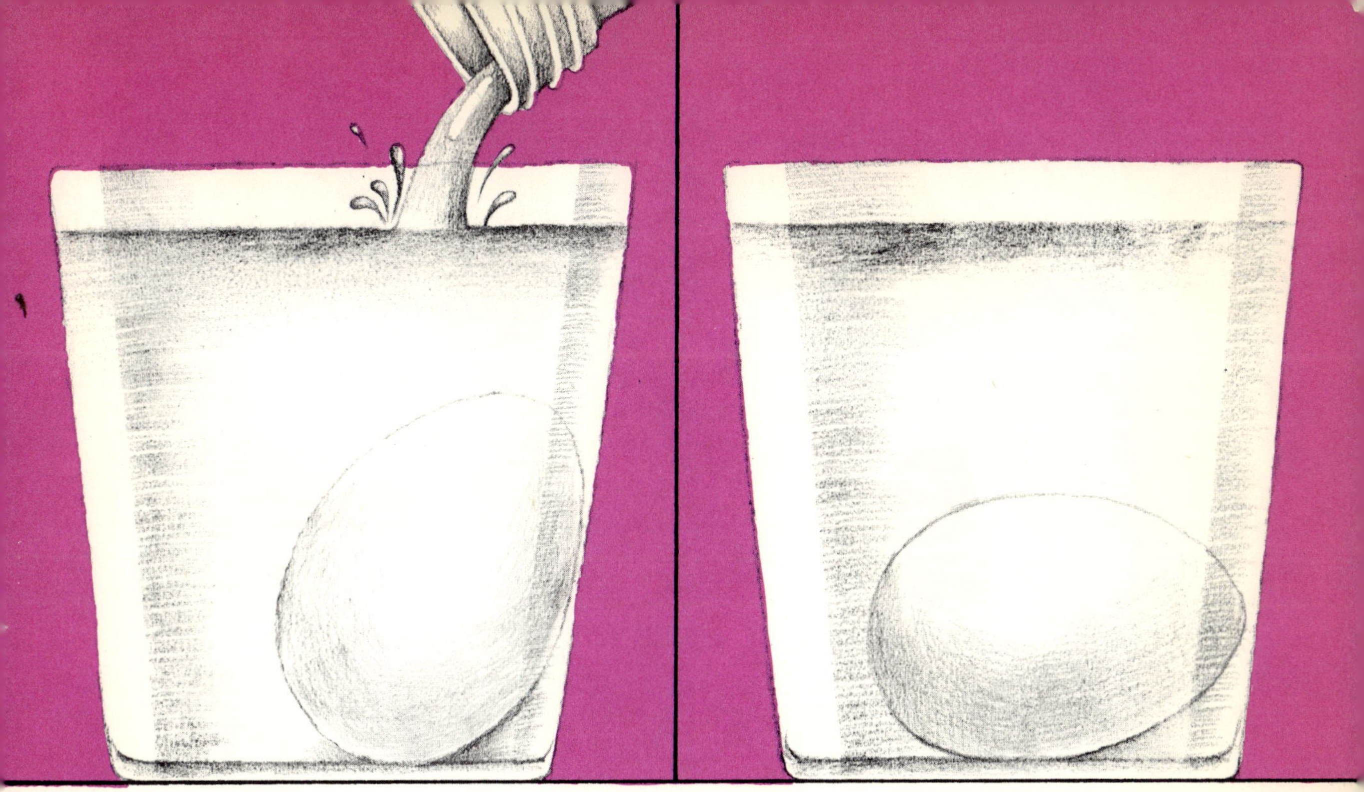

Clean a Dirty Penny with Chemicals

Put two spoonfuls of vinegar in a cup.
Sprinkle in some salt. (Five shakes will do.)
Drop in a dirty penny.
Stir it around.
Watch!
See how it shines when you wash it off?

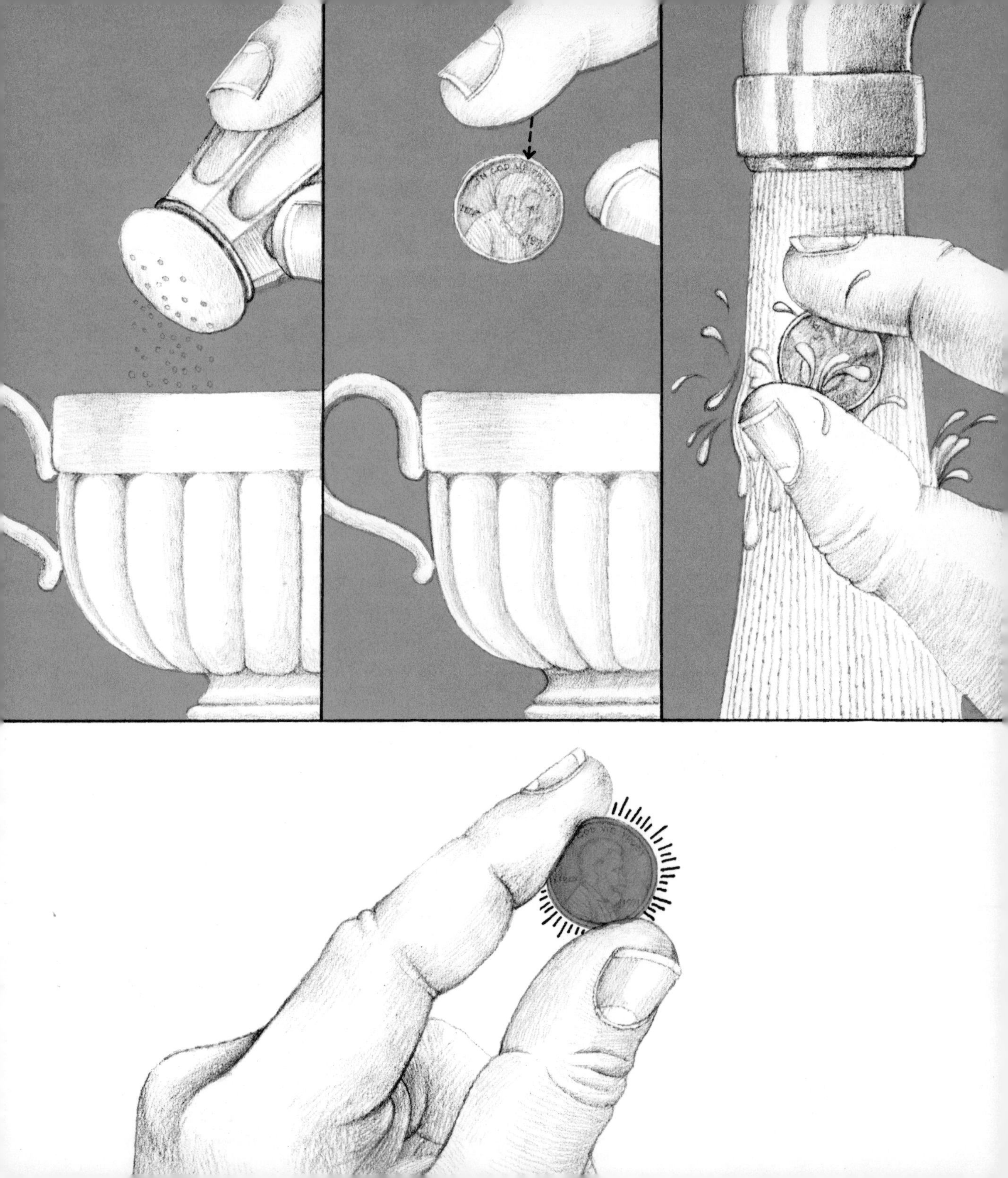

See Water in a Potato

Scoop a hole in a raw potato.
Use a spoon.
Put a spoonful of salt in the hole.
Leave it until tomorrow.
Tomorrow the hole will be full of water.
There is a lot of water in a potato.
Salt makes some of it come out.

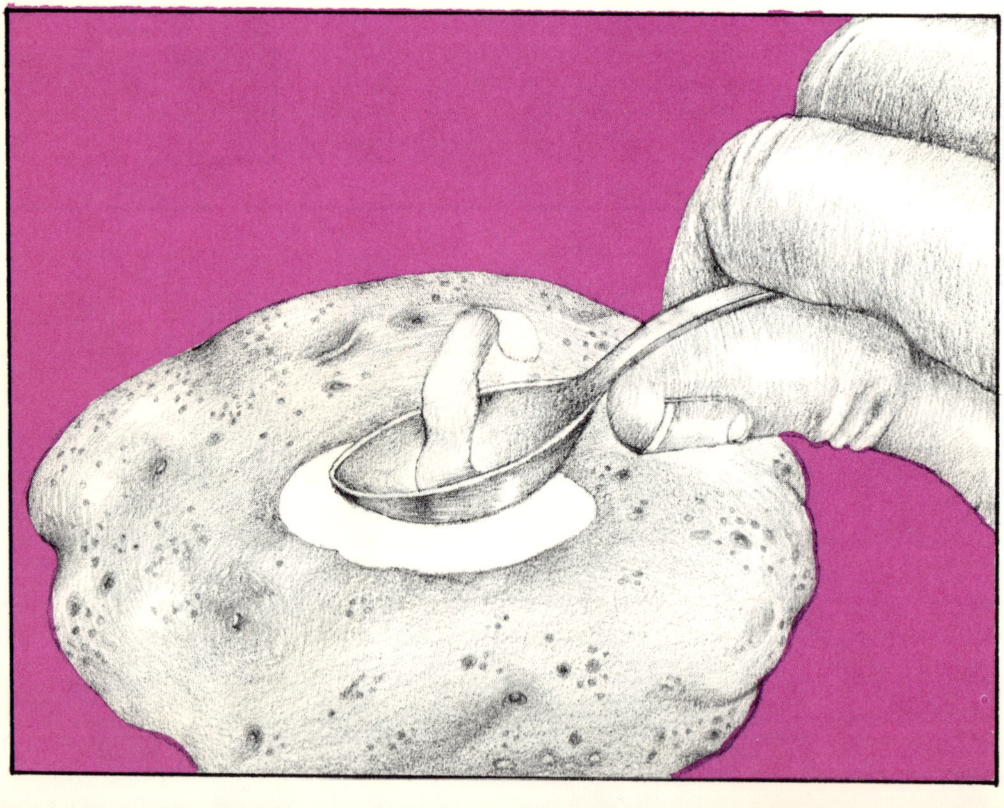

An Ice Cube Turns Over by Itself

Put an ice cube in a bowl of warm water.
Do not touch it.
It will turn over all by itself.
Keep watching.
It will turn over again.
And again.
Why?
The warm water melts the bottom of the ice quickly.
The top of the ice gets heavy.
The top falls and the cube turns over.
And over, and over, and over.

Feel How Hot Your Hand Is

Open your hand out flat.
Hold it close to your face.
Leave a little space between your hand and face.
Hold still.
Your body is warm all the time.
Your warm face and hand make the air warm, too.

Find a Place You Cannot See

There is a place you cannot see.
It is right in front of your face.
Look at your finger.
Can you see it?
Slowly move it to the top of your nose.
Point right between your eyes.
The end of your finger will be gone.
You cannot see the top of your nose.
You cannot see anything that is put there.

See One Big Eye

You have two eyes.
They both see the same things.
Put your nose on a friend's nose.
Push your faces close together.
What does your friend look like?
Both his eyes look like one.

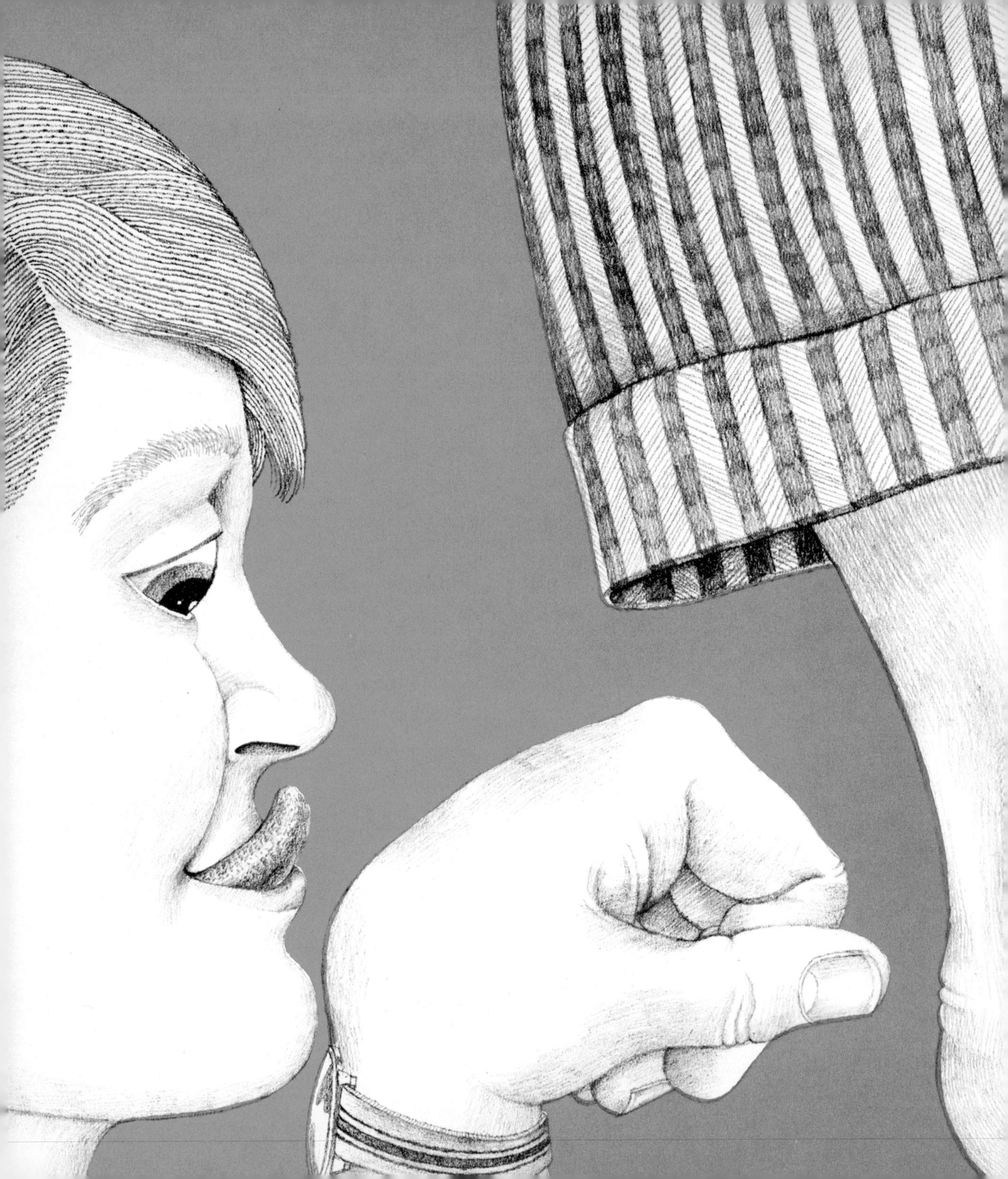

Most People Cannot Do These Things

Here are three things to try right now.
You may not be able to do any of them.
Touch your fingers to your wrist.
Touch your tongue to your nose.
Make *just* your little toe wiggle.
Your body can do so many things.
It is a surprise to find things you cannot do.

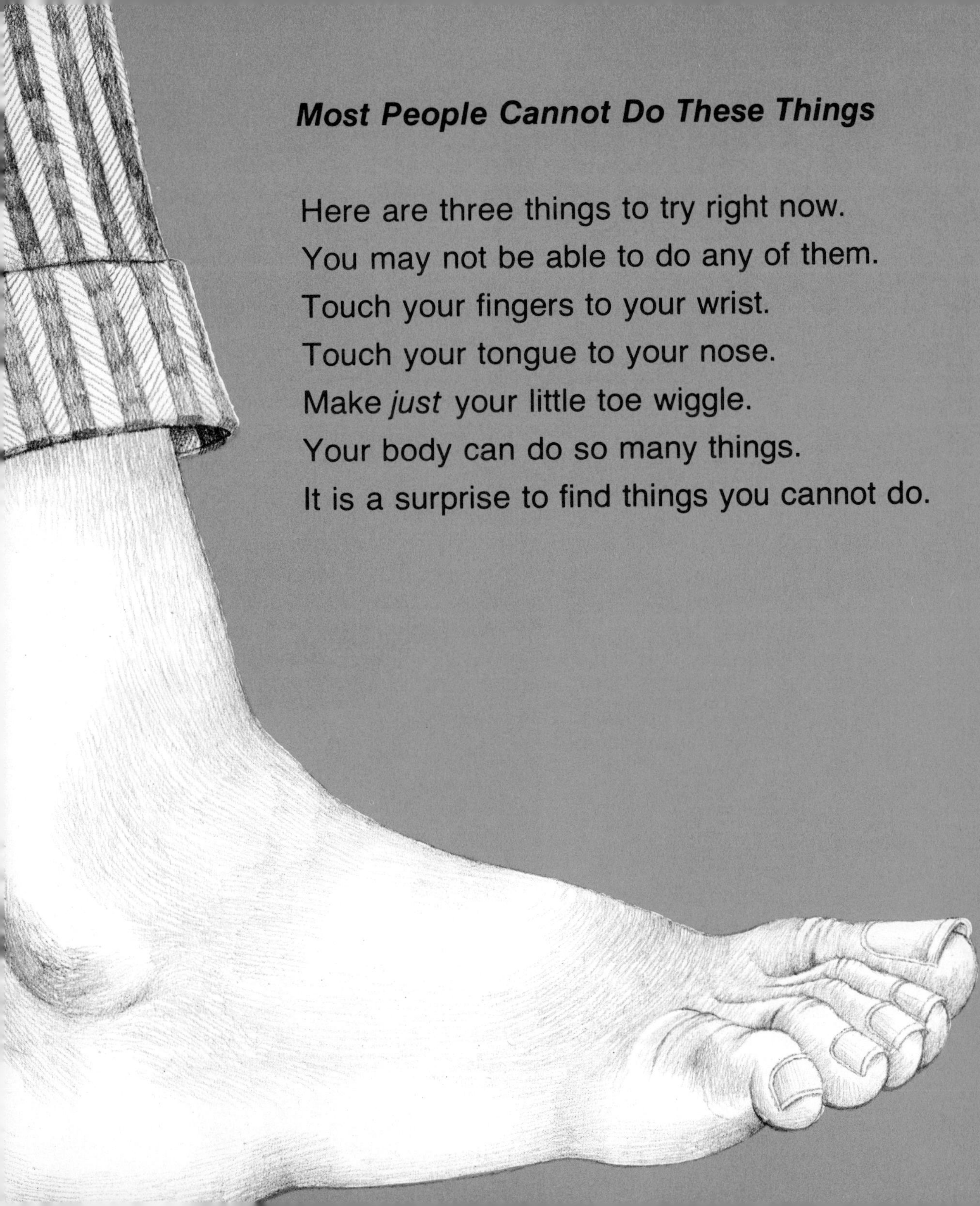

Try to Touch Your Toes

Stand with your back up against a wall.

Try to touch your toes, without bending your knees.

You will not be able to do it.

Your body moves forward.

You fall over every time.

Can You Blow This Tent Over?

Cut a piece of paper about this big.
Fold it in half.
Open it like a little tent.
Set it on a table.
Try to blow it over.
You will not be able to do it.
Air holds it on the table.

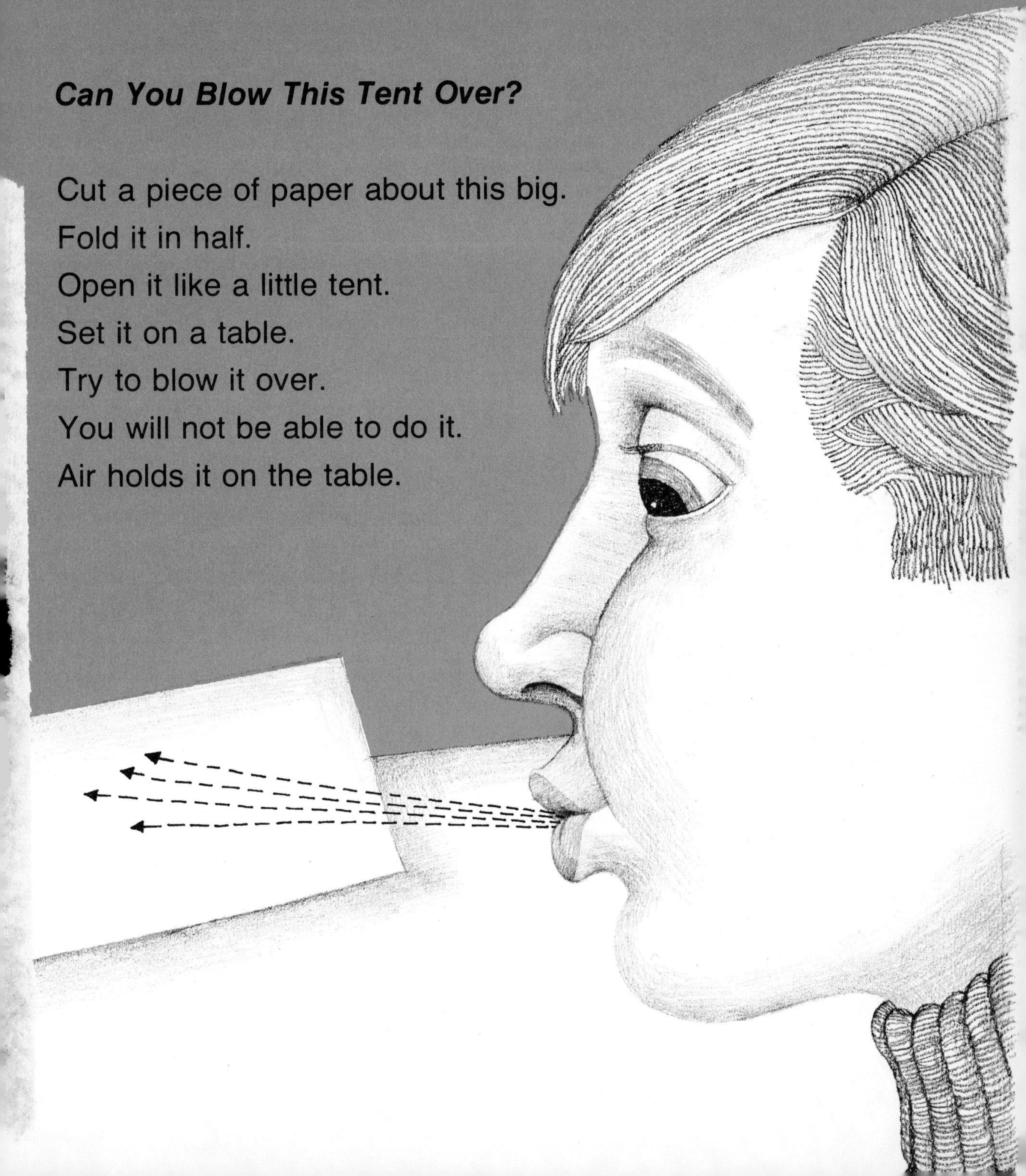

Blow a Piece of Paper Up

Hold a piece of paper by two corners.
Let the end hang down.
Put your lips close to the paper.
Blow hard over the top.
The paper will go up.
Why?
Your breath is like the wind.
The paper is like an airplane wing.

A Sheet of Paper Can Hold Up a Book

Take a sheet of paper.
Then take a book.
Can the paper hold up the book?
Roll the paper into a tube.
Put a rubber band around it.
Set the book on it.
Tubes and pipes are very strong.
They are much stronger than flat things.

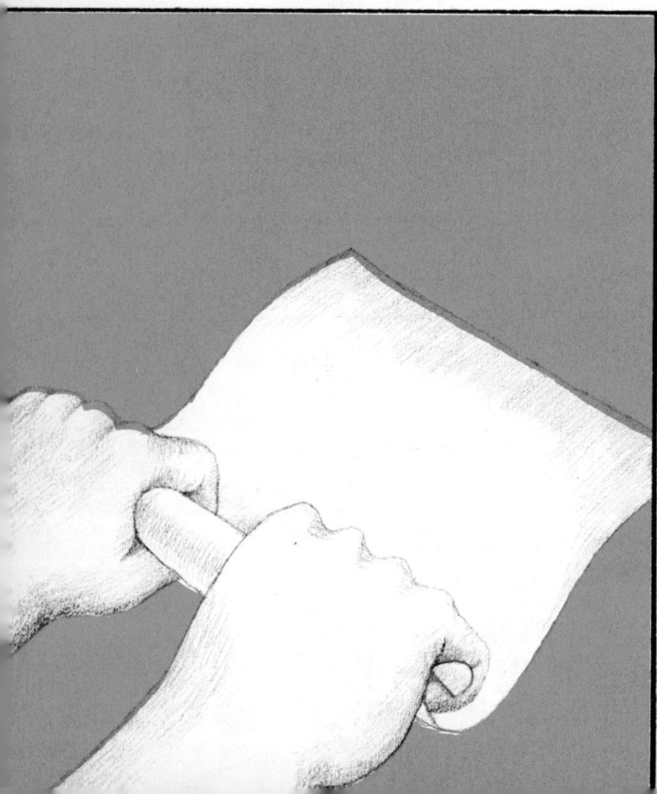

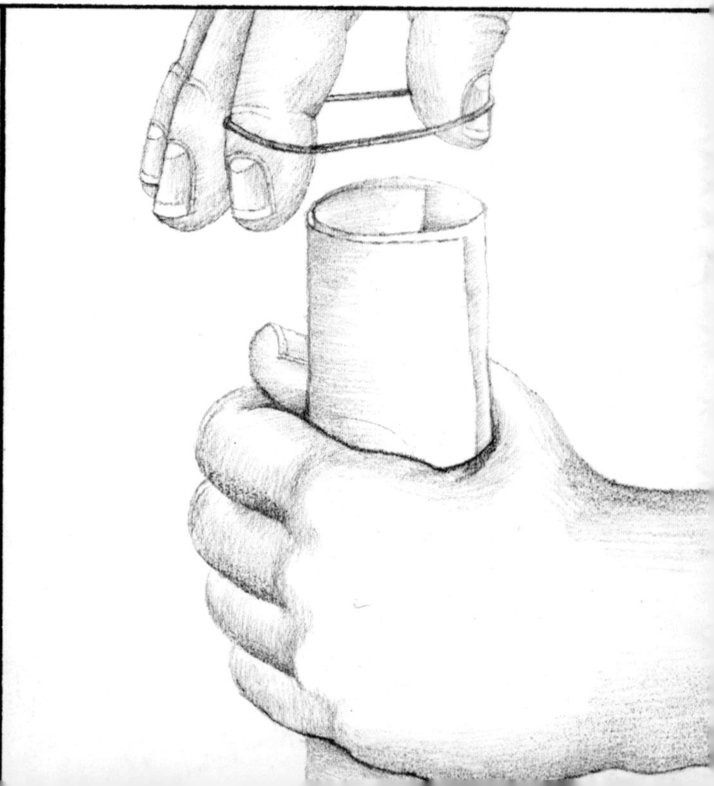

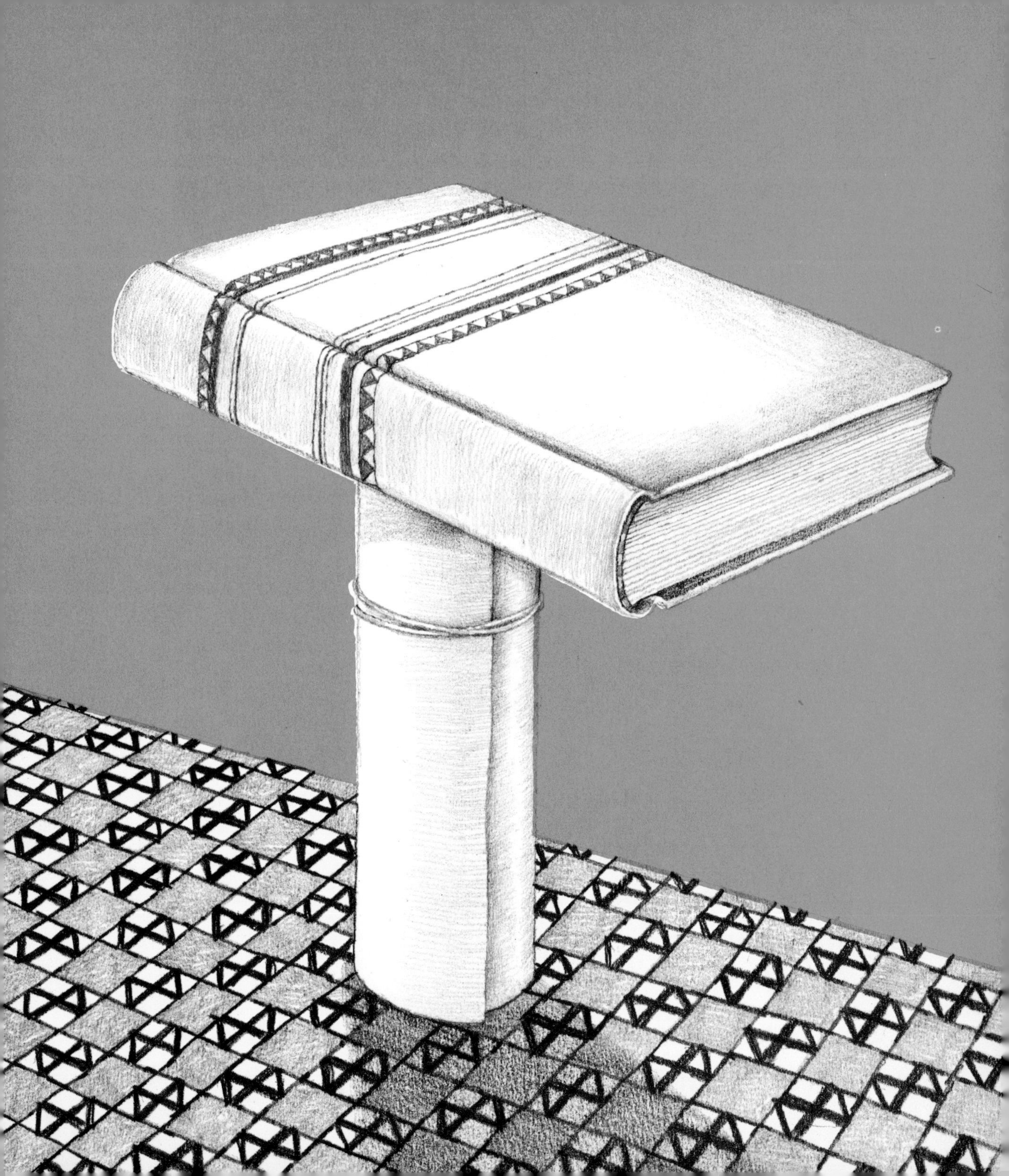

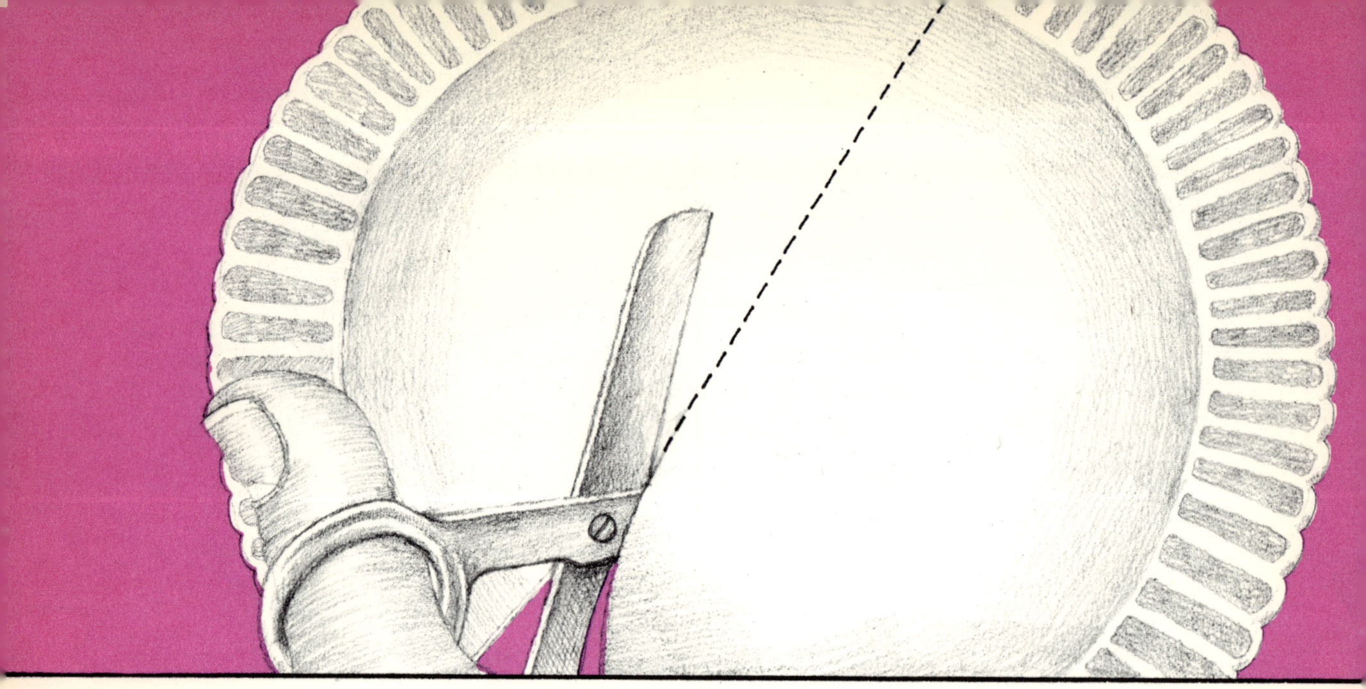

Fool Your Eyes with an Optical Illusion

Cut a paper plate in half.
Cut the middle out of one half.
Cut the curve in half.
Put one piece over the other.
Cut them both the same length.
Hold them side by side.
One looks longer.
But it isn't.
It is an optical illusion.
An optical illusion is a trick that fools your eyes.

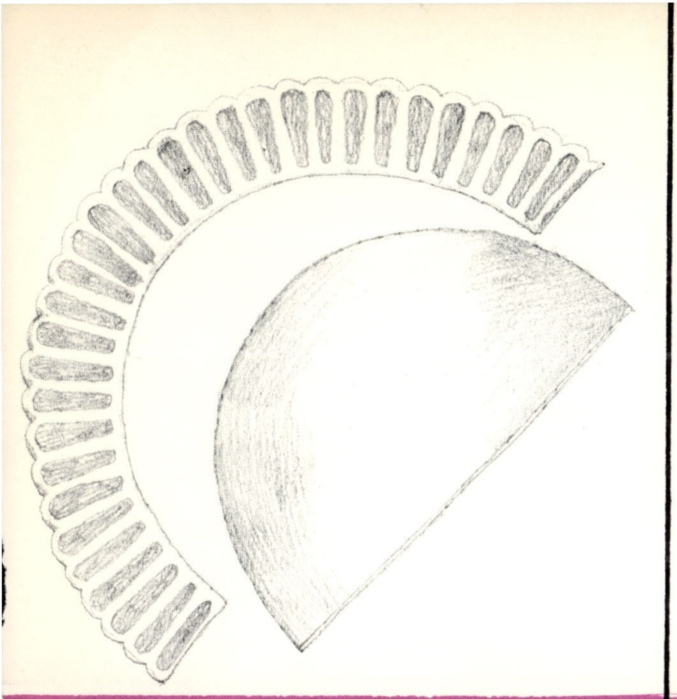

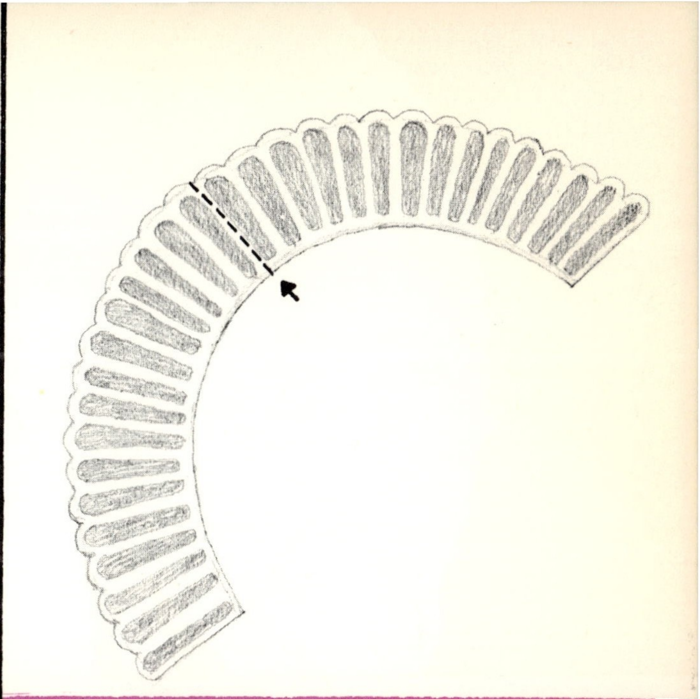

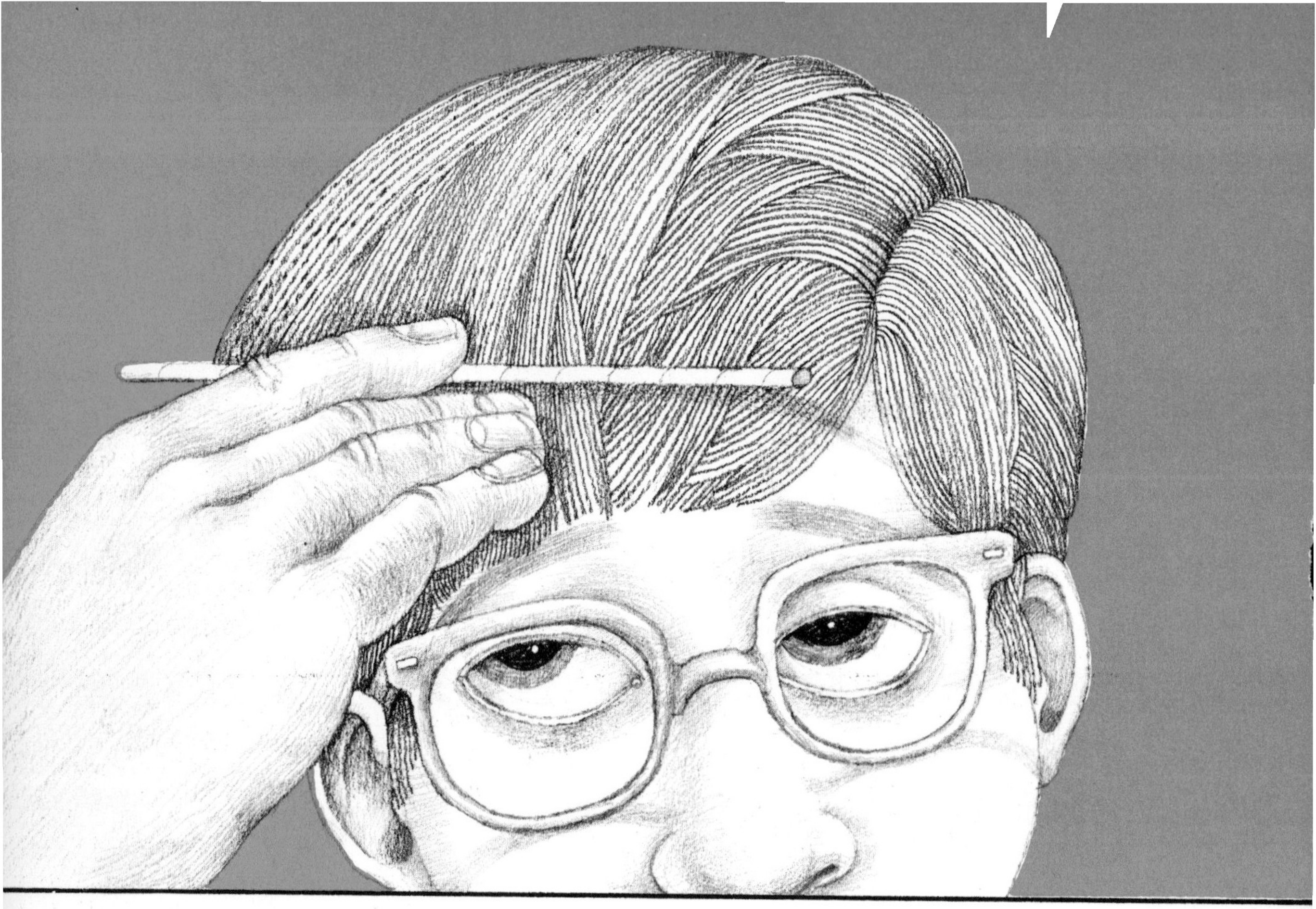

Make a Straw Move with Your Hair

Lay three plastic straws on a table.
Rub another straw in your hair.
Move it near the top straw.
The top straw will roll to it.
Static electricity makes the straw roll.
You made static electricity with your hair.

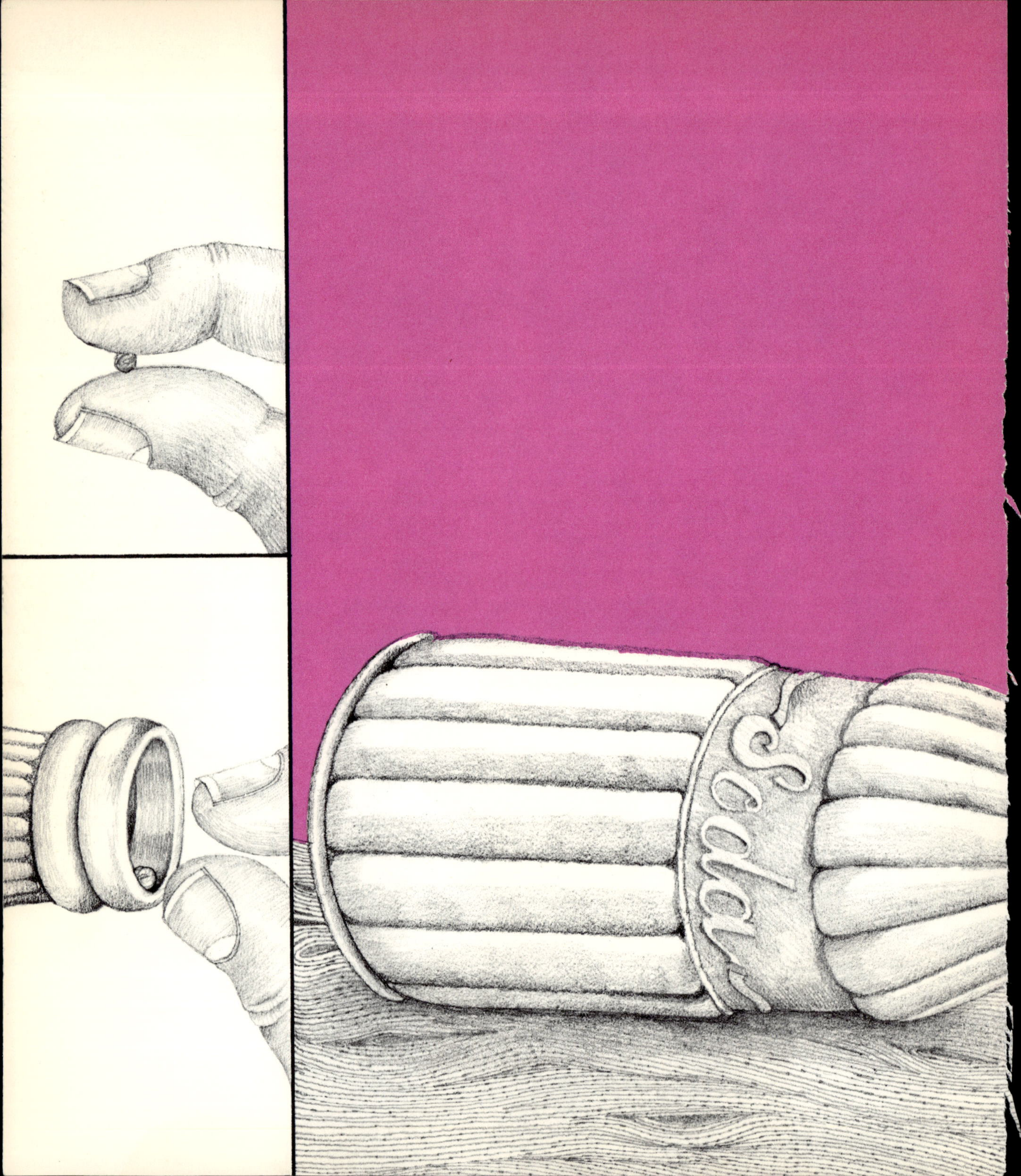

Blow In to Blow a Ball Out

Lay an empty soda bottle on a table.
Roll a little piece of paper into a tiny ball.
Put the ball just inside the bottle.
Try to blow the ball into the bottle.
The ball will pop out.
The bottle is full of air.
The air pushes the ball out.

There are no more tricks to try
in this book.
But there are many more
in the world around you.
See how many you can change
from magic into science.